Ecology provides a fascinating introduction to the ways plants and animals interact, both with one another and with their environment. The world's major habitats are described and illustrated, and man's effects on the natural world are discussed in simple terms.

Acknowledgments:
The publishers would like to thank Wendy Body for acting as reading level consultant and Chris Tydeman, who works for the World Wide Fund For Nature, for advising on scientific content.

Artwork credit:
Illustrations on pages 10-11, 34-35, 38-39 and 43 by John Dillow.

Photographic credits:
Pages 17 and 30, Ardea/I. R. Beames; pages 13, 14 and 37, Ardea/F. Gohier; page 19, Ardea/C. Haagner; page 25 (top), Ardea/S. Meyers; page 24, Ardea/P. Morris; page 25 (bottom), Ardea/Swedburg Photo; page 29, Greenpeace/Cannon; page 27, Greenpeace/Merjenburgh; page 36, ICCE/T. Snook; page 9, Photri/Robert Harding Picture Library; page 42, back cover, World Wide Fund For Nature/D. Reid; page 31, World Wide Fund For Nature/ P. M. Snyder.
Designed by Anne Matthews.

British Library Cataloguing in Publication Data

Ganeri, Anita
    Ecology.
    1. Ecology
    I. Title   II. Woolf, Colin
    574.5
    ISBN 0-7214-1230-0

First edition

Published by Ladybird Books Ltd Loughborough Leicestershire UK
Ladybird Books Inc Auburn Maine 04210 USA
© LADYBIRD BOOKS LTD MCMXCI

Printed in England (3)

Ecology

written by ANITA GANERI
illustrated by COLIN WOOLF

Ladybird Books

What is ecology?

Animals and plants live in many different places – rain forests, mountains, deserts, lakes and oceans. The study of where and how animals and plants live together is called ecology.

A rock pool is home to many plants and animals. Fish and tiny shrimps shelter among the seaweeds. Starfish eat mussels which cling to the rocks. Hermit crabs use old shells to protect their soft bodies.

All plants and animals are linked closely to one another and to their surroundings. Our surroundings and everything in them are called the environment.

Any changes in the environment affect the animals and plants that live there.

Homes and food

There are lots of special places in the environment where particular animals and plants live. Each special place is called a **habitat**.

This picture shows a **food chain**.

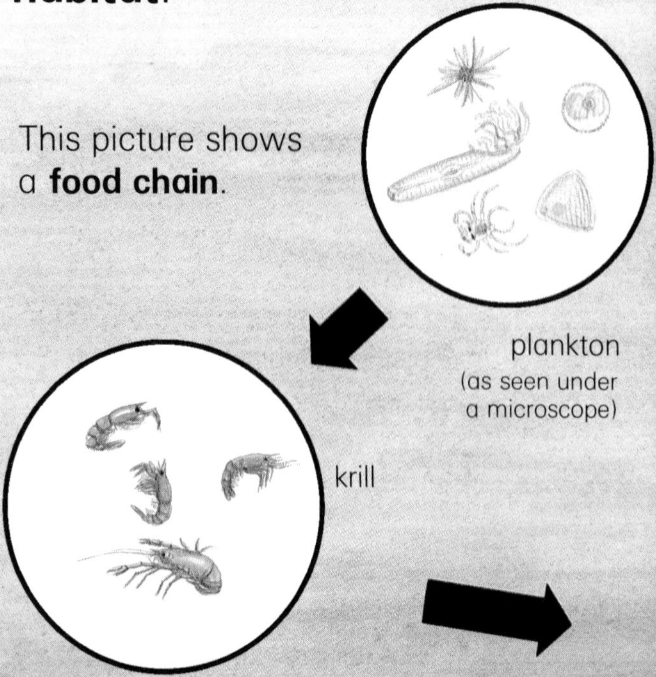

plankton
(as seen under a microscope)

krill

The plants and animals that live in a particular habitat are called a community.

In icy Antarctic waters, shrimp-like creatures called krill eat plankton (tiny plants and animals). In turn, krill are eaten by fish, which in turn may be eaten by penguins. The animals and plants in this community are all part of a food chain.

penguin

fish

Changes in nature

The environment is continually changing. Usually the changes happen slowly over a long time, perhaps millions of years.

Other changes, like volcanic eruptions or man-made disasters such as oil spills, have sudden and dreadful effects.

The Earth's climate changes naturally but slowly over a period of time. During the last Ice Age, 19,000 years ago, many kinds of animals died out because it was so cold. Others grew thick, shaggy coats to keep them warm.

In 1980 Mount St Helens, a volcano in the USA, erupted and a large part of the mountain was blown away. Over 500 sq km of forest were flattened.

mammoth

Map of habitats

This map shows the world's main habitats. Many plants and animals have special features to

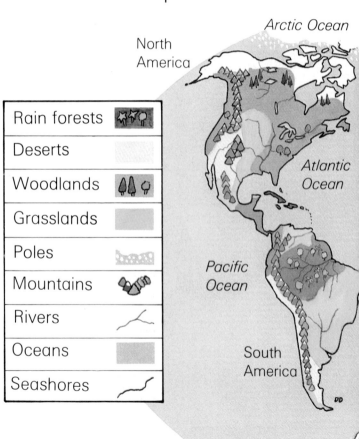

Rain forests	
Deserts	
Woodlands	
Grasslands	
Poles	
Mountains	
Rivers	
Oceans	
Seashores	

Arctic Ocean

North America

Atlantic Ocean

Pacific Ocean

South America

help them to survive in their particular habitat. This is known as **adaptation**. For example, desert lizards called geckos have webbed feet to stop them sinking in the sand.

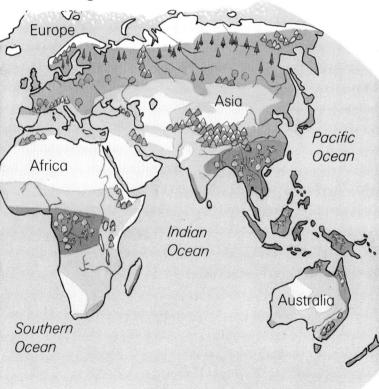

Europe

Asia

Pacific Ocean

Africa

Indian Ocean

Australia

Southern Ocean

Antarctica

Tropical rain forests

Two-thirds of all the known plant and animal **species** in the world live in hot, wet **tropical** rain forests. These forests provide wood, rubber, medicines and fruit for people.

Arrow-poison frogs, woolly monkeys and toucans live high among the rain forest's trees, where bright orchids and ferns grow. On the forest floor, tapirs and snakes hunt for food.

Today rain forest habitats are in great danger. So many trees are being cut down that there may be no forest left in 40 years' time.

Destroyed rain forest in Central America.

Deserts

Deserts cover about a fifth of the Earth's land surface. In the daytime, most deserts are very hot, but they can be freezing cold at night. Even though there is very little rain, many plants and animals live there.

Owls and other animals shelter in holes in cactus stems to escape the fierce daytime heat.

The Sahara Desert is expanding south by about 1 km a month. This is because too many animals and crops have been raised on the land at the edge of the desert. The soil there has become very dry and has blown away.

Addax antelopes
and kangaroo
rats never
drink water.
They get the liquid
they need from
eating plants
and seeds.

Woodlands

There are two types of **temperate** woodlands on Earth. Huge forests of fir and pine trees grow in the colder north. Woods of beech and oak trees grow in warmer places.

These forests are homes to thousands of different kinds of animals.

Today people are clearing woodlands and building homes on them. Some animals, such as foxes, learn to survive in towns.

Other animals cannot adapt. They die when their natural homes are destroyed.

Acid rain destroys woodlands. It is formed when poisonous chemicals from factories rise into the air and mix with rain and snow.

Grasslands

Grasslands are huge, dry areas of grass with scrubby bushes and a few tall trees. They are found in many parts of the world and are home to hundreds of different animal species.

In some African countries, huge areas of grassland have been turned into national parks. Here the animals are protected from hunters.

The Etosha National Park in Namibia is one of the best places to see African elephants in the wild.

The poles

The Arctic and Antarctic are the coldest places on Earth. The Arctic Ocean is frozen all year round. Thick ice covers the land at the Antarctic.

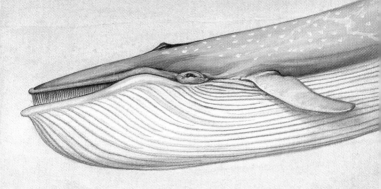

Gigantic blue whales live in the cold Antarctic seas. In the 1930s there were about 200,000 blue whales in the world. But so many were killed for their meat and fat that at one time their numbers dropped to just 2,000.

Many countries
have now agreed
not to catch whales
until 1996.

Seals, penguins and many other animals
survive because they have thick,
waterproof coats and special layers of fat
under their skin to keep them warm.

Mountains

High up in the mountains fierce winds blow and it is freezing cold. Life is difficult for animals and plants. But mountain slopes are home to goats, marmots, snow leopards and condors. They all have thick fur or feathers to keep them warm.

No trees can grow near the tops of very high mountains because it is too cold. An imaginary line called the tree line marks this area on the mountain.

In winter some animals,
such as marmots, **hibernate**
to escape the cold.
Others, like wild sheep
and ibex, move farther
down the mountain to find grass
to eat. They are followed and
hunted by hungry snow leopards.

23

Rivers

Most rivers start as fast-flowing mountain springs. As they flow towards the sea, they get wider and slower.

Tiny caddis fly larvae live in the swirling upper river. They spin silky webs to catch food as it is carried down the river.

Many river animals, such as hippos and alligators, prefer to live in slow water.

Many river habitats are being destroyed by pollution. Poisonous waste from factories kills plants, insects and fish, which other animals need for food. Pollution may also destroy the water's oxygen supply, which the animals need to live.

Along the seashore

Along the coast the sea rises and falls with the tides. Seashore animals and plants have to find ways of keeping wet when the tide is out.

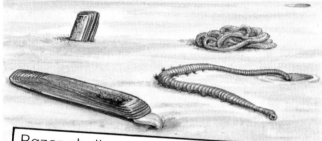

Razor shells and lugworms burrow into the damp sand when the tide is out.

Seaweeds have slimy coats so they do not dry up when the tide is low.

Sea anemones pull in their tentacles and sit like blobs of jelly on the rocks. When the tide comes in they unfurl their tentacles, ready to catch food.

In March 1989 the *Exxon Valdez* oil tanker spilt 45 million litres of oil in the sea near Alaska. The oil soon drifted ashore, polluting the beaches and bays. Many thousands of sea birds and fish died.

Oceans

The oceans cover nearly three quarters of the Earth. They are the world's biggest habitat and home to thousands of animals and plants.

Coral reefs are only a small part of the ocean habitat, but they contain about one third of all the world's species of fish. The Great Barrier Reef near Australia is so large it can be seen from the moon.

Each year thousands of dolphins get tangled up in nets used to catch tuna and squid. The trapped dolphins cannot reach the surface to breathe, so they drown.

Animals in danger

All over the world animals are in danger of becoming **extinct**.

Many lose their homes as land is cleared for building and farming and the seas and rivers are polluted. Seals and leopards are killed for their fur coats.

The tiny golden lion tamarin from Brazil is one of the most **endangered** animals. Some zoos are breeding tamarins to release them back into the wild.

Rare parrots and spiders are caught and sold as pets. Today, many threatened animals are strictly protected. It is against the law to catch or kill them.

African elephants are being killed for their ivory tusks, which are made into jewellery and ornaments. Governments around the world have now banned this trade.

The greenhouse effect

The glass in a greenhouse traps the Sun's heat and keeps the plants warm. Carbon dioxide gas in the Earth's **atmosphere** does a similar job and keeps the Earth warm. This is called the greenhouse effect. But factory chimneys and car exhausts are producing too much carbon dioxide and the Earth is in danger of overheating.

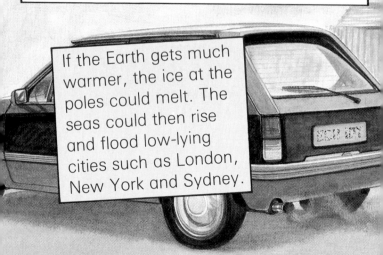

If the Earth gets much warmer, the ice at the poles could melt. The seas could then rise and flood low-lying cities such as London, New York and Sydney.

To stop the Earth getting warmer, we need to stop burning so much wood, coal and oil. Burning these fuels produces huge amounts of carbon dioxide.

The ozone layer

About 12 km above the Earth is a layer of gas called ozone. This acts like a screen, protecting us from the Sun's harmful ultraviolet rays. In 1985 scientists discovered a hole in the ozone layer above

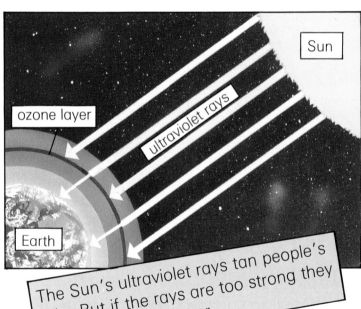

The Sun's ultraviolet rays tan people's skin. But if the rays are too strong they can cause skin cancer.

Antarctica. They think that chemicals called CFCs (chlorofluorocarbons) may be destroying the ozone layer. CFCs are found in aerosol sprays, polystyrene packaging and refrigerators.

To help to save the ozone layer, we need to stop putting CFCs into the air. Look out for CFC-free labels on hairspray and deodorant aerosols and plastic food trays in the supermarket.

Saving energy

We use huge amounts of **energy** for heating, cooking and to power cars and factories. Many people use **fossil fuels** – coal, oil and natural gas – as their main sources of energy.

Our supply of fossil fuels will not last for ever. In some countries, people are now using the Sun's energy instead. They fit special solar panels to their houses. These convert the Sun's energy into electricity.

But these can pollute the air.
Nuclear power is also used but
it too can be very dangerous.
Scientists are experimenting
with cleaner, less harmful
sources of energy.

We could also use wind power in the
future. The Altamont Pass wind farm in
California, USA, has 300 wind turbines.
As the wind turns the blades, electricity is
produced.

Recycling rubbish

We each throw away about 100 dustbins of rubbish every year. This includes glass bottles, tin cans and paper. We then buy new things which have taken a huge amount of

It takes much less energy to recycle objects than to make brand new ones. Recycling aluminium cans, for example, takes twenty times less energy than making new ones.

energy to make. A lot of this rubbish could be used again, or **recycled**. This saves energy and cuts down on the number of trees and other materials we need.

You can help to save energy by collecting your used bottles, soft drink cans and newspapers. Many towns have special 'banks' where you can take them to be collected for recycling. Find out where yours are.

Saving the land

Only about a tenth of the land on Earth is used for farming. But each year more crops are needed as the population grows. If farmers use a piece of land too much, the soil loses its minerals and becomes too dry

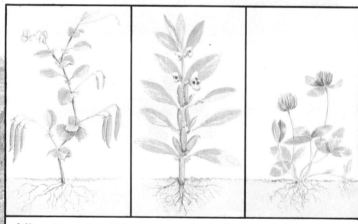

Nitrates are important for plant growth. Most crops take nitrates out of the soil. But peas, beans and clover put them back in. Many farmers grow these in their fields every few years. This is called crop rotation.

to use. Today some farmers are trying to protect the soil and still get bigger and better harvests.

Farmers must protect their crops from harmful pests. Chemical sprays called pesticides kill insects but also poison the birds that eat the insects.

Today, more farmers are starting to grow organic crops, which are free from man-made chemicals.

The future

Scientists believe that one plant or animal species becomes extinct every half hour. Once a

We all need to act now to save the Earth and its wildlife. This is called **conservation**. Conservation groups such as the World Wide Fund For Nature, Greenpeace and Friends of the Earth are already working hard to save wildlife. They have helped to protect tigers, whales and pandas.

species dies out, it can never be replaced. Elephants and gorillas, for example, may disappear for ever if we continue to kill them and destroy their homes.

You can help to look after the environment by picking up litter from the street and countryside. You could also learn more about the plants and animals in your garden and talk to other people about caring for our world.

Glossary

adaptation A change in a plant or animal that results in its becoming better suited to its environment.

atmosphere The thick blanket of air which surrounds the Earth.

conservation Protecting and saving wildlife and habitats.

endangered Animals and plants which are in danger of dying out for ever. Some examples are giant pandas, African elephants and golden lion tamarins.

energy The fuel needed to make things work. This may be food energy, heat energy, light energy and so on.

extinct Animals and plants that have died out and no longer exist on Earth.

food chain A series of plants and animals linked by their feeding habits, each being eaten by a larger one that in turn feeds a still larger one.

fossil fuels Coal, oil, natural gas and peat are fossil fuels. They are made from the remains of dead prehistoric plants and animals.

habitat The type of place where a plant or animal lives. Deserts, mountains and oceans are all types of habitat.

hibernate To spend the winter in a resting state. Many animals hibernate, including bears, tortoises and hedgehogs.

recycling Collecting and converting rubbish into useful things.

species A group of animals or plants which are very alike and can breed with one another.

temperate Weather that is hot in the summer, cold in the winter and moderate in the spring and autumn.

tropical Weather that is hot and usually humid all year round.